The Beekeeping Year

Lynfa Davies,
Master Beekeeper, NDB

The Beekeeping Year

by Lynfa Davies

ISBN: 978-1-914934-79-7

Published 2024 by Northern Bee Books, Scout Bottom Farm, Mytholmroyd, Hebden Bridge HX7 5JS (UK). 01422 882751.

Book design by www.SiPat.co.uk.

Contents

Introduction

Beekeeping is a fascinating and rewarding hobby, but for many new beekeepers it can also be daunting. While for the most part, bees are very good at looking after themselves, their welfare and productivity will be a direct result of your actions. How do you know what to do and when? And how do you know what to expect?

This book is divided into monthly sections to give you an idea of the sequence of events throughout the year and how to plan for them. For example, when is swarm control most likely to be required and what equipment should be prepared? Another important topic, covered in April, is what to look for during inspections. The health of your bees is paramount, and this is covered throughout the book to help you learn what to look for and the actions you need to take.

There is always something to do no matter what time of year it is and understanding what time commitments you need for your new hobby will help you to plan. Coupled with this is learning how the colony grows in the spring, peaks in the summer and contracts in the autumn. We call this the colony cycle and there are things we need to do to help our colonies respond to these natural rhythms and reach their full potential.

I hope this book helps to set the scene for you and guide you through the beekeeping year. You will not learn everything in the first season and mistakes will be made but keep going and everything will start to make sense. If I can give you three top tips to help you get started it would be;

- Join your local beekeeping association (their help will be invaluable)
- Plan ahead (even simple things like having equipment ready will make a difference at busy times)
- Don't panic! (there will always be help and advice available when you are just not sure what is going on).

Enjoy your new hobby and the fascinating insight into the life of your bees that comes with it. I hope you gain as much from it as I have.

1.
The Apiary in January

Practical jobs with the bees at this time of year may be few and far between, but there is still plenty you can be doing. Hopefully you have had a fruitful Christmas and you can spend these long winter nights reading the books you were given and improve your knowledge of beekeeping. There is always something new to learn and as you accumulate this knowledge, you will be surprised to find how it does help with decision making in the apiary. You may even be inspired towards taking one of the assessments on offer, either practical or theoretical, and the quiet winter months provide an ideal time to do some of the background reading required. Studying as a group is generally far more productive, particularly if you have been out of formal education for a while. It is worth approaching your association to see if they can facilitate this.

The bee cluster

Out in the apiary life will be fairly quiet at this time of year. The bees form a compact cluster when the ambient temperatures drop below 14°C and this cluster tightens to maintain the warmth as the temperatures drop. Once the temperatures fall below freezing the bees will start to generate heat rather than tighten the cluster further. The outer shell of bees is relatively calm and quiet, with the heads facing inwards and they form the outer shield preventing too much heat loss. Meanwhile, the bees on the inside are more active because they are warmer. The bees will also rotate positions, in much the same way that the penguins in Antarctica all take their turn on the outer edge of the group.

↑ Snow does not present a problem to the bees, but clear
blocked entrances in prolonged snowy periods.

Isolation starvation

Periodically, the bees will need to break the cluster and allow workers to move to where the feed is stored. The bees may starve if they become separated from their feed stores and it is too cold for them to break away from the cluster. This is difficult to protect against, so the best advice I can give is to ensure your bees have plenty of stores going into the winter (20-25kg) and if you are in any doubt at this time of year, offer them some fondant. Place this under the crown board over where the bees are clustered, so that they don't have to break from the cluster to take advantage of it. Keep the fondant wrapped in plastic with only the side adjacent to the bees open, so that it doesn't dry out. Filling old takeaway tubs with fondant works well for this and then use a shallow eke between the top of the brood box and the crown board.

Monitoring food resources

Hefting your hives is a useful skill to develop so that you can try and get a feel for how much of the stores have been consumed. Start by hefting the hives immediately after you have finished the autumn feeding. Lift one side of the hive and make a mental note of how heavy it feels and then move to the other side and do the same. You can repeat the exercise at various times throughout the winter and compare how it feels compared to the first time you hefted. I have to say this is an art and not a science! And do beware of the woodwork of the hive taking on moisture during the winter months, which will make it heavier. If you find the hive is losing weight, it may be time to offer the bees some fondant as an insurance policy.

Varroa

The varroa mite continues to be a threat to the health and well-being of our bees and mid-winter is a good time to treat. In the winter oxalic acid based treatments are used, either by the trickling method or by sublimation. This method relies on a broodless period as this ensures that the mites are on the adult bees rather than being hidden away in sealed brood cells. Recent research shows that the broodless period is much earlier than previously thought making early December the optimum time to perform a winter treatment with oxalic acid and this is described in more detail in December. However, if you have not yet done it you can still do so, but the science shows that if there is brood present, you should open any brood by scratching the surface of the cappings away. Beekeepers are often concerned about opening their colonies and disturbing them at this time of year, but it is a relatively quick job because there is so little brood, and the bees quickly settle down again. Choose a mild, calm day and check the middle of the box first as you probably won't need to go through the whole colony.

← Peel back the plastic on the fondant on the underside and place directly over the cluster. An eke has been added here to lift the crown board away from the top of the brood box and leave space for the fondant.

Security

If you keep your bees in your garden regular checks to make sure they haven't blown over are easy. But, if like me, you have out apiaries, you will need to make regular visits to check all is well. Strap your hives down in the winter, particularly those away from home and this will avoid late night panics when an unexpected storm blows in off the Atlantic! Occasionally there are reports of badgers knocking over hives in the winter when their food supplies are getting scarce, and straps will also help stop their pilfering attempts. Green woodpeckers can also be a problem during the winter when the ground is too hard for them to get at the ant nests that they target. A cage made from chicken wire placed over the hive, will stop the woodpeckers gripping the side of the hive and therefore they cannot position themselves to drill into the side of the hive.

While heavy snow is less common these days, especially where I live, we do occasionally get caught out. Snow does not present a problem to the bees and they will simply respond by clustering tighter until the snows have gone. If you do encounter a prolonged period of snowy weather just check that the hive entrances are not blocked as you will be surprised how often bees take cleansing flights during the winter on bright, sunny days. It is also worth reassessing the position of your apiary in the winter. At this time of year trees and hedges will be bare of leaves and may make the hives vulnerable to winds that may not cause a problem when the full hedgerows can protect them. Cold, easterly winds are likely to be the most damaging so do make sure your hives are protected against them.

2.
The Apiary in February

The first spring flowers will be bursting into life now which is a welcome sight for our bees on mild, sunny days. Crocuses and snowdrops will provide much needed fresh pollen and signs of this being taken into the hive is a good indication that your queen has started laying. Pollen is an essential component of brood food and it is unlikely that there will be any left in the brood nest from last autumn. The old winter bees will use their internal fat reserves and any pollen they can gather to produce the brood food for the larvae. If you want to promote the spring build-up of the hive consider feeding pollen substitutes which are fed in the same way as fondant. Also, remember to heft the hives to provide an indication of stores in the hive and if you gave fondant last month check again to see whether they have used it.

Prepare equipment

The new beekeeping season is fast approaching so now is a good time to start preparing your equipment. We are all guilty of putting away equipment at the end of the season without cleaning it or doing any repairs so make the most of this relatively quiet month and get your scraper and blowtorch out. Scrape off any brace or burr comb and propolis and collect these scrapings into a bucket – don't leave them where bees can get at them in case they are harbouring disease. Wooden supers and brood boxes, floors, crown boards and rooves can be sterilised with a blowtorch. The idea is to pass the flame over the woodwork just enough to lightly singe it – you are not trying to set it alight! This is very effective at destroying potential pathogens such as nosema and can also get at wax moth larvae hiding in the corners. The stainless steel queen excluders can be blowtorched but do not use a blowtorch on the zinc slotted or plastic excluders as these will warp.

↑ **Snowdrops are a welcome sight in February for us and the bees.**

Polystyrene hives can be cleaned by scrubbing with a strong solution of washing soda (1kg soda crystals to 5 litres water) after gently scraping off any propolis and wax. If you want to sterilise them, you'll need a plastic box large enough to submerge them in. Fill this with a solution of bleach (1 part household bleach to 5 parts water) and then immerse the hive parts for at least 20 minutes. They will need to be weighted down as they will float.

Sterilise combs

Empty, stored combs can be sterilised by using 80% acetic acid which is particularly effective against nosema and chalkbrood. Make a stack of the boxes of combs on a solid floor and place a saucer of acetic acid (120ml per box of frames to be sterilised) on the top using an empty super as an eke. Place a solid lid on top and seal up any gaps with parcel tape and leave for a week. Acetic acid is corrosive to concrete, metal parts and humans so remove any metal bits, avoid spilling on concrete and wear the appropriate protective equipment. After a week open up and allow to air for a few days before using.

When fumigating comb, I use heavy duty plastic boxes that have perforated sides and are stackable. The combs are suspended in the boxes and absorbent towels are placed on top. I pour the acetic acid onto these towels and then put the next box on top. They are all wrapped up in large plastic silage bags for the duration of the fumigation process. This can also be done by simply stacking the wooden brood or super boxes and applying the acetic acid on absorbent pads to each box before sealing up.

Set some new goals

Taking the time to plan some goals for the forthcoming season is a very worthwhile exercise at this time of year. In my early beekeeping days, I was never particularly organised and my activities were a reaction to what was going on at the time. For example, I never thought about swarm control until I found some queen cells and then I would have to react to this. This is all very well, but once you have a few colonies it can become rather exhausting and it feels like the bees are in control rather than you! Now I think about what I would like to achieve over the season and then prepare accordingly. Here are a few suggestions to illustrate what I mean and no doubt you can think of others that you would like to do.

Prepare to take a practical assessment. I have no doubt in my mind that the practical assessments available through the beekeeping organisations are a good way to improve your skills. Start with the Basic and then as you become more experienced you can work towards the Honey Bee Health certificate, the General and Advanced Husbandry assessments and the Bee Breeding Certificate. Speak to your local training education officer and be aware that closing dates for these assessments often fall in February!

More effective swarm control. This is a priority for all beekeepers, so take the time to read up on this topic and select a method which suits you. It may be that you would like to try some pre-emptive swarm control by making splits. This may help you manage the bees better without losing half of your strongest colonies. But you will end up with twice as many colonies, so think about how this will affect you – do you have time and equipment for more colonies? How

← Preparing combs for sterilisation. This can be done in the wooden brood boxes or super boxes but do not use polystyrene boxes.

about trying out a Snelgrove board? This method can result in high stacks of boxes which is not suitable for someone who does their beekeeping alone. But the method can be adapted and used 'horizontally' thus avoiding high stacks of heavy boxes. Wally Shaw describes this method in his Apiary Guide to Swarm Control.

Rear your own queens. This is a very rewarding process and doesn't need to be as complicated as some make out! A little research and planning will reward you with a method to try. Don't be put off if it's not totally successful on the first attempt – it may just be a case of more practice is needed and with a little more experience you will be able to refine the technique.

My advice is don't make your list too long. Pick a couple of things at a time and over time these will all add up to make you a better overall beekeeper and make this fascinating hobby even more rewarding.

3.
The Apiary in March

March is full of anticipation and excitement for what the forthcoming beekeeping season will bring. Weather patterns over recent years have brought mild, sunny conditions in March only to be followed by a cold snap later in the spring. Some of us may be able to do our first spring inspections in March while others will still be in the depths of winter. Don't rush into anything. As long as the bees have sufficient stores all should be well.

Feeding

Feeding is a real balancing act at this time of year. If their stores have run out, you may need to feed because the queen will be laying and the brood nest will be expanding, but the bees may not be able to forage enough to support this growth. Be careful to avoid over-feeding your colonies, as this will cause growth that is too rapid. This results in colonies swarming as soon as the weather warms up because they've filled their brood nest! If the weather is warm enough for the bees to break their cluster you can offer a 1:1 syrup but offer fondant if the weather is still too cold.

If you are targeting the oilseed rape crop you will need to be stimulating the colonies with some feed to ensure you have a sufficient foraging force by mid-April. If you are concerned that not enough pollen is coming in for brood rearing, you can also offer pollen substitutes which are fed the same way as fondant.

First inspection

We are always keen to do our first spring inspection and see how the colony has fared over the winter. Choose a warm, sunny day of approximately 15°C before opening the hive. Don't spend too long in there – have a quick look for evidence of a laying queen and check that they have stores. It is worth counting the number of frames of brood in the brood nest as this will give a valuable comparison the next time you inspect. The size of the brood nest should be increasing at this time of year, so at the next inspection you should expect to see one or two more frames of brood in the brood nest. If you over-winter with a super below the brood, swap them back around and if you are sure your queen is in the brood box, you can place a queen excluder between the two boxes. Don't worry if the weather conditions don't allow this until early to mid-April as there will be huge variation across the country.

↑ A colony that has died from starvation. Notice how many bees are headfirst in the cells.

Colony losses

Unfortunately, we occasionally lose colonies over the winter. This is disheartening to the beekeeper, who naturally feels they have done something wrong. In 2021 the BBKA Annual Colony Survival Survey reported losses in the UK of 18.6% over the previous winter, with beekeepers suggesting various reasons why they thought their colonies had failed, including; queen failure, starvation, lack of forage and cold weather. This figure varies every year but remains relatively high. Attention to management in the summer and autumn, particularly varroa control, will help control losses. If one of your colonies dies, close it up immediately to prevent other bees accessing it until you are able to clean and sterilise it. Identifying and understanding why a colony has died is not always easy but the following pointers may help:

- Bees are clustered together and some are head-first in the cells. This is probably starvation and is characterised by a lack of stores. Sometimes the bees get separated from their stores and they may run out of food in the combs surrounding them even though there are stores in other parts of the hive. This is called isolation starvation and occurs when it is too cold for the bees to break from the cluster and travel across the comb to their stores. This is best avoided by ensuring that colonies are strong and have plenty of stores at the start of winter.

- A lack of brood and/or evidence of raised cappings in worker brood comb. The number of bees has dwindled. This is likely to be queen failure, with the raised cappings caused by a drone laying queen.

- Dead bees with deformed wings and dried out pupae that have been nibbled. This is parasitic mite syndrome caused by varroa and highlights the importance of timely, effective mite control.

- Evidence of faeces on the frames and at the hive entrance, along with low numbers of bees. This may be more evident during the spring build up, as these colonies stand out from the healthy ones as they do not build up. This is a sign of Nosema and can be controlled through good hygiene practices and comb management.

- Always check dead colonies for evidence of disease, particularly American Foulbrood (AFB). This bacterial disease kills the developing larvae after the cell has been capped. The result is a dark hard scale that is stuck to the lower side of the cell that cannot be removed by the bees. If you suspect you have AFB scales in your hive close it up immediately to prevent other bees accessing any stores left in the hive and call your bee inspector.

Prepare frames

Make sure you have enough frames for the season ahead. There are different types of frames and spacing options available and it is important to get the spacing correct to make it easier to work your colony. Hoffman frames are self-spacing because of the design of the side bars of the frames. When they are pushed together in the hive the frames are held at the correct spacing. There are alternative options including frames with straight side bars which need to have a spacer added to the lug. You can buy plastic end spacers for this, but they do become clogged with propolis. Manley frames offer another alternative for your supers and they have wider top and side bars. If the spacing is incorrect the bees build brace comb to fill gaps that are too wide. If you see brace comb being built between inspections have a look at the spacing in the hive – the problem may be easily solved. Castellations are another form of spacing which are best used in the supers. I know some people like to use castellations in their brood boxes but this means you cannot slide the frames when you are doing hive inspections. Sliding frames in groups or all together is a useful technique to avoid squashing bees and you cannot do this if they are in their fixed positions in the castellations. The options available may seem a bit confusing and the best thing to do is to have a look at what is available in a beekeeping supplies catalogue and then speak to a more experienced beekeeper about which is likely to suit you best.

← This colony dwindled and died out due to a drone laying queen.

4.
The Apiary in April

April can be a fickle month with very changeable weather. While some will still be waiting to do their first inspection others will be reporting the first swarms of the season! Don't be impatient – you have to work with your local conditions, and you will be amazed how quickly you tune into these as you gain experience with your bees. If you do manage to inspect your bees the things you need to be checking are:

* Stores
* Space
* Health
* Queenright
* Queen cells

Stores

A cold snap may interrupt the foraging activities of your bees and may even delay the early spring flowers. Keep an eye on food reserves at this time of year because the brood nest will be rapidly expanding and very vulnerable to a check in food supplies. You can supplement your bees with some fondant and pollen substitutes if they have run out of stores, or even a little light syrup if the weather is warm enough (1kg sugar to 1 litre of water) using a contact feeder.

↑ Healthy sealed brood has an even dry appearance, with domed cappings that are biscuit coloured.

Space

On the other hand, the weather may be lovely and warm and the flowers out in abundance. If this is the case, and let's hope so, then you need to have supers ready to ensure the rapidly expanding colony has space to store the nectar they are bringing in. As a rule of thumb, once the bees are covering 8 frames it is time to give them space by adding another box.

Do not forget that the queen may need additional space to lay, and you may need to consider adding a second brood box (referred to as double brood) or allow her to lay in a super by placing one below the queen excluder (brood and a half).

↑ A typical pattern on the comb showing the arc of brood surrounded by pollen and then stores in the top corners.

When adding new supers the question often asked is do you place the new super on top of the existing supers (top-supering) or below them closest to the brood nest (bottom-supering)? This always causes much debate with the theory suggesting that the bees will work the new super quicker if it is closest to the warmth of the brood nest. However, there appears to be little difference. If your colony is not strong enough to move into the super and start working it, regardless of where it is positioned, then the colony probably does not need a new super!

Health

Spring is a good time to do a disease inspection. Ideally you should aim to do 3 or 4 disease inspections throughout the active beekeeping season and with a little practice you can give the colony that all important health check. Choose a warm day when the bees are flying well as you will have the hive open a little longer than usual. First find your queen and pop her in a queen cage but remember to return her to the colony when you have finished. Starting with the first frame of brood shake all the bees from the frame to give you a clear view of the brood. The important thing here is to be able to recognise what healthy brood looks like. It should have an even laying pattern and the cappings will be dome shaped and biscuit coloured. Uncapped brood will be pearly white, lying in a C-shaped position in the bottom of the cell, with clear segmentation. Once you are happy with identifying healthy brood you can start being more critical and look thoroughly for signs of disease. This could include dead larvae that are brown in colour or even hard and chalky. Are the cappings looking sunken and greasy or maybe the laying pattern is uneven, and some larvae have been removed from their cells? Hold the frame at a 45° angle with the sun shining over your shoulder and check for signs of American Foulbrood scales in the bottom of the cells. Space on this page does not allow full blown descriptions of all the diseases you could encounter, but do not panic as further advice can be found in the National Bee Unit's (NBU) leaflet on spring checks and if you do suspect one of the foulbroods, close the hive up, reduce the entrance to prevent robbing and contact your bee inspector.

Queenright

During your inspection check that you have a queen that is laying well and the colony is building up as expected. Marking your queen is a good idea to help you spot her, but if you cannot find her, check that you have eggs in the brood nest. These should be laid in rugby ball shaped patches in the middle of the brood nest and as she fills the nest, the eggs will be in concentric arcs gradually moving out from the centre of the nest. Record how many frames you have at each inspection and then you will be able to see if the brood nest is expanding. You would expect the brood nest to be full of brood by the end of May, so in April it will be reassuring to see a gradual increase towards this. If there is no increase week on week it could be down to a number of reasons, but the most likely causes are lack of food, disease or a failing queen.

Queen cells

Depending on how well advanced the season is and the size of the colony, don't forget to check for the onset of swarming by keeping an eye out for queen cells. These will most likely be around the edges of the frames and can be well hidden. Be prepared by planning your swarm control method in advance and having all the necessary equipment to hand.

The oilseed rape crop

The oilseed rape will start flowering in early April and if you are taking hives to the rape you need to do so before the main flow starts. Remember to plan your journey, however short it is, and make sure all the hives are securely strapped down and well ventilated for the journey. Take plenty of supers with you so that you can give the bees plenty of space. Many of you will be taking advantage of the oilseed rape from your home apiaries but you still need to be prepared and ensure the colonies have plenty of supers plus spares in case there is a good flow.

Asian hornet traps

Now is a good time to put up traps for Asian hornets (*Vespa velutina*) in your apiary. We all need to be vigilant to ensure that any Asian hornets are discovered as early as possible and reported to prevent them becoming established here. The NBU provides details of a trap that can be made from a plastic bottle and hung in the apiary. Alternatively, you can also purchase traps from beekeeping suppliers. Understandably, there is concern about the number of non-target insects caught in these traps so please check them regularly to release these. The important thing is that we all play a role in monitoring and any suspect hornets must be reported to alertnonnative@ceh.ac.uk

5.
The Apiary in May

Activity in the apiary should now be in full swing. It is an exciting time with rapidly expanding colonies taking advantage of the spring flowers. Have you prepared enough equipment for the forthcoming season – well, you will soon find out!

Weekly inspections

Weekly inspections will be required so that you can monitor how well the colony is doing and keep a watch for signs of swarming. In April, I described what to be looking out for during an inspection and the same applies as you move into the active season. Keeping records at each inspection is an essential thing to do and can be done in a variety of ways. Some of us will prefer a simple paper method and record observations such as: level of stores, number of frames with brood, signs of swarming and health. Others may prefer digital methods, but the same information is recorded. If you are planning to rear your own queens it is worth recording other pieces of information such as: temper, calmness on the comb, honey yields and varroa levels. The list of things you can record is endless and it really depends on what sparks your interest as to how much detail you want to record, but the important thing is to record the basics.

I use this information to monitor how well a colony is building up. Have the number of frames with brood increased over the last few weeks? If not, why not? Is it lack of forage, a queen that is not laying very well, or is there an underlying health issue? Armed with the information I am collecting I can take steps to put things right. This could include putting in the inspection boards to monitor varroa or giving a light feed if foraging activity is low. Piecing all the information together will help you build up a picture that enables you to manage the colonies and this will be made much easier and more effective if based on records.

Comb management

May is a good time of the year to think about comb management. The aim is to remove old, dark, well used comb that could be harbouring disease and replace with foundation to enable the bees to produce new comb. The long-winded way is to gradually move the old dark comb to the edge of the brood nest and then, when it is clear of brood, remove it and replace with foundation in the centre of the brood nest. This can take a while but does work. The alternative is to undertake a Bailey Comb Change or a Shook Swarm. These have the advantage of speeding up the process and bees respond really well to their 'clean sheets', but it does require some careful management to avoid stressing the bees.

For either of these manipulations wait until the weather conditions are favourable and there is a honey flow. Bees need warmth and food to draw comb. A honey flow can be artificially provided but it is harder to create the warm conditions!

Bailey Comb Change

A Bailey Comb Change involves placing a brood box containing frames of foundation over the existing brood box. A temporary entrance is placed between the boxes to reduce traffic through the old comb. The queen is moved to the top box on a frame of comb and a queen excluder placed between the two boxes. The colony is fed and once all the old brood has emerged from the bottom brood box it can be removed and cleaned up and the hive reconfigured. This is an excellent way to get them onto new comb. It doesn't involve destroying brood, so there is no check to the colony development. However, it is a relatively slow method (3-4 weeks) and if there are disease pathogens present, they will be transferred to the new comb. So, this is a useful method as long as you are confident there are no disease issues.

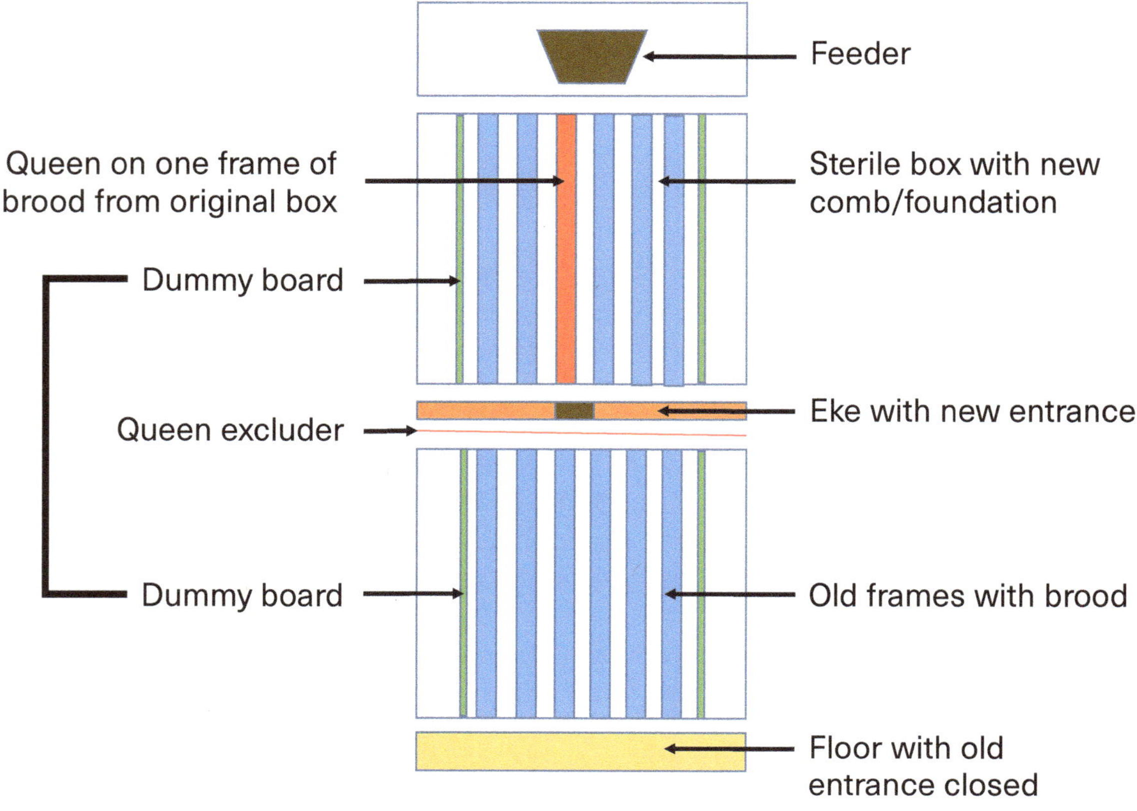

↑ This diagram shows how to set up the colony for a Bailey comb change. The queen is in the top box with one frame of brood and new combs or foundation. The remaining brood is below the queen excluder where the brood will emerge and then the old frames can be removed and the colony put back together again. Remember to give them a feed to encourage them to draw new comb.

Shook Swarm

The Shook Swarm method appears a little brutal because you immediately remove all the existing brood, but in my experience the bees respond really well. It also has the advantage of delaying swarm preparations in that colony. For this method move the existing brood box to one side and in its place put a queen excluder on the floor and place the new brood box containing frames of foundation on top. Remove a few frames from the centre to create a bit of space. Find the queen in the old brood box and pop her in a queen cage during the manipulation. Now shake all the bees from the old comb into the gap in the new brood box, run in the queen and fill the gap with the remaining frames. Add a feeder and provide them with some light syrup, replace the roof and leave for a few days. When you return you will witness one of nature's miracles – comb building. I am always amazed how quickly they draw out the comb and it looks so fresh and clean when it is first built. Once you see eggs and brood in the new comb you can remove the queen excluder that is under the brood box. This was there to prevent the colony absconding, which they might have done because you took them from their lovely warm brood nest and shook them into an empty box!

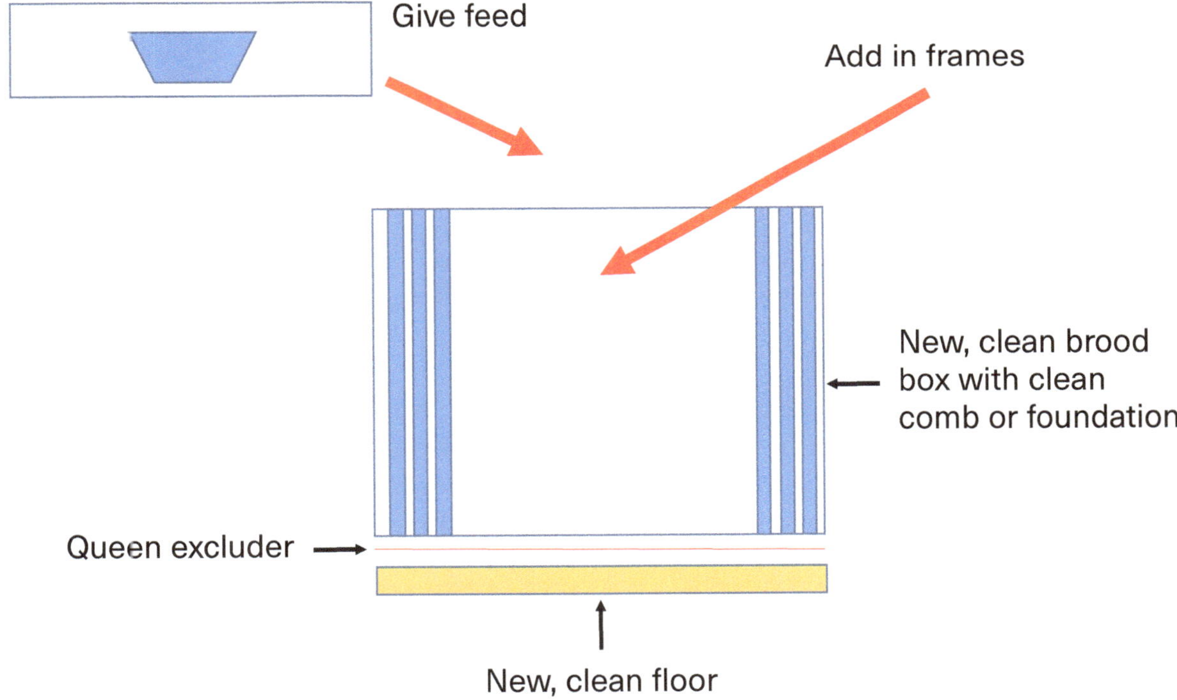

↑ This diagram shows how to do a shook swarm. The gap between the combs allows you to shake the bees from the old frames into the clean box. Then you add the rest of the new frames and give them a feed. They will quickly draw new comb to give their queen somewhere to lay.

↑ A frame from a colony 2 weeks after a shook swarm. The whole brood box was drawn with a brood nest covering 6 frames in the centre.

Not only is the Shook Swarm a useful way of delaying the onset of swarming, but it is also an excellent way to reduce varroa levels early in the season. The majority of the varroa mites in a colony are safely hidden away within the sealed brood cells and feeding on the developing bees. When you remove all the old comb and brood, you are removing most of the varroa from that colony. This is a win-win for the bees and, as long as you help them along with a bit of food, they will respond and build up to full strength very quickly. Now you have your bees in good health and at full strength – bring on the summer!!!

6.
The Apiary in June

June is a lovely time of year and a busy one for beekeepers. It is likely that at some point, you will need to manage your colony's impulse to swarm. There is no disgrace in having a colony that is making swarming preparations – this is perfectly normal. It is what bees do to reproduce their colony and set up a new home. As beekeepers we will spend much of our time trying to reduce or delay this impulse until we are forced to take action. There are good reasons why we should control swarming and try to prevent it:

- Loss of honey crop if your colony is suddenly halved in size
- Spread of disease including varroa
- Public nuisance - not everyone loves bees as much as we do!

Despite extensive research, no single trigger has been identified as a cause of swarming. In reality, a combination of factors contribute in some part, and trying to manage all of these factors is the challenge. Let's look at what we can do.

- Provide space in the brood nest and supers. Overcrowding is one of the triggers that can lead to swarming. Provide the queen with plenty of space to lay which may mean adding a second brood box or a super (for a brood and a half configuration). Don't forget to add supers ahead of them needing one. Bees need space to process nectar and they also need space to hang out and rest. When you have a large colony, there will not be sufficient space in the brood box to accommodate them.
- Make weekly inspections to check for signs of queen cells that indicate the bees are preparing to swarm. Take appropriate action if they are found.
- Use rapidly expanding colonies (as these are the ones most likely to swarm) to make splits. This will not prevent swarming, but it will delay it.

If you find unsealed queen cells during your inspection, it is time to act. There are several methods of swarm control that can be employed, and it is worth reading up on these well before the season starts, so that you are prepared. There are many useful resources available on this topic.

↑ **Managing swarming is important particularly when you have a narrow gauge train station full of tourists at the bottom of your garden!**

A Spring Honey Harvest

In a good season, where the spring was warm and forage was abundant, you may be lucky enough to have a full super of honey that is worth extracting. Spring honey comes predominantly from trees such as hawthorn, fruit trees and sycamore. It tends to be darker in colour than summer honeys and is well worth extracting so that you can enjoy its flavour. Also, if your bees have been working oilseed rape you need to extract those supers promptly. Oilseed rape honey will granulate very quickly so you need to extract it before this happens otherwise you cannot spin it out from the frames.

You may have heard references to a 'June Gap'. This is a natural break in nectar availability when the spring flowers go over but before the summer ones are available in any abundance. It can be a problem for large colonies because they will be unable to find sufficient forage to maintain themselves and, if you have already removed a spring crop, they may very quickly run short of stores. It does not occur every year but you do need to be aware of it. During your weekly inspections, make a note of how much stores the bees have and if it is depleting rather than increasing you may need to consider a light feed to tide them over. This is a bit of a juggling act as you do not want them to store syrup in their supers, but equally you do not want the colony to have a check as this could impact on their ability to bring in a summer crop. As always, close monitoring and recording of what you see is useful to make comparisons week on week.

Varroa

Keep an eye on varroa levels at this time of the year. It is easy to think that the colony is doing well as it is bursting at the seams, but lurking in the depths are those pesky varroa mites happily breeding in some of the brood cells, particularly the drone brood. Use open mesh floors with the inspection tray in for approximately a week and then count the number of mites on the tray and divide by the number of days the inspection tray was used. This gives the daily mite drop and figures above 10 indicate that treatment is required.

The temptation is to avoid treatment while there are supers in place, but this can be a mistake if varroa numbers are building rapidly, and it could result in the colony collapsing very suddenly in mid-late summer. Full details on the available treatments can be found on BeeBase but alternatively, there are options for controlling varroa that do not involve chemical treatments and should form part of your Integrated Pest Management.

Drone brood removal can be employed for this purpose. If a brood frame is replaced with a drawn super frame, the bees will nearly always build drone comb in the gap below the super frame. Once the queen has laid this up and the bees have sealed the comb, cut it off and throw it away. It will contain a significant number of the colony's varroa mites as they prefer to breed in drone brood. Make sure you do remove the capped drone brood if you are using this method and do not leave the drones to emerge, otherwise you will release a fresh flush of mites into the colony and make matters worse! Alternatively, you can insert a frame split into three parts and set it up to have a section ready for removal each week. There are other effective methods of varroa control without using chemical treatments and full descriptions can be found on BeeBase in the booklet titled 'Managing varroa'.

↑ Drone brood removal on a rolling weekly cycle.

Check for disease

Remember to check for other signs of disease in your colonies at this time of year and June is a good time to check for diseases such as European Foulbrood and chalkbrood in the same way as described in April.

7.
The Apiary in July

The Welsh word for July is 'Gorffennaf' which literally means the 'end of summer' and since becoming a beekeeper I have realised how true this is. As a child you think of July as the start of summer because this is when your holidays begin, but for a beekeeper, July is the time to start thinking about winter preparations. Of course, many of us will still be getting nectar flows, dealing with late swarms, and preparing for the heather, but do not let these activities distract you from the fact you need strong, healthy colonies to get through the next winter.

Continue to monitor varroa levels and treat if necessary. The temptation is to wait until after honey has been removed, and rightly so, as some treatments must not be used with honey supers in situ. However, if the colony is already reaching breaking point, there is evidence of deformed wing virus and mites are obvious on the bees, I would recommend addressing this now to safeguard the colony. Refer to the National Bee Unit's advice in their booklet 'Managing varroa' for details of the different treatments and how to use them.

↑ Rosebay willowherb is a common and important food source for the bees at this time of year.

Harvest Honey

At this time of year, the bees will be reliant on wildflowers as most arable crops will have finished flowering. Bramble and rosebay willowherb are the mainstay for many of us in Wales and if the weather is kind can give good honey crops. Your equipment stocks can become a little stretched at this time of year as all the hives are stacked with supers. Extract the honey as soon as it is capped and then you can re-use the supers if nectar is still being brought in. Use clearer boards under the full supers to remove the bees and, if possible, give them an empty super under the clearer board to provide them with some space. There are several different methods available for clearing bees from supers most of which rely on the principle of a one-way valve i.e. bees leave the supers but cannot return to them. None are 100% effective, but any remaining bees can easily be brushed off the frames. It is best to remove supers from the apiary in the evening when the bees have stopped flying, otherwise you risk being swamped with enthusiastic helpers all trying to regain access to your supers!

Protect against robbing

Remember that once the supers are empty of bees, they are very vulnerable to robbers. By this I mean opportunistic bees from other colonies who have found a quick and easy source of food. Robber bees can be a real nuisance and because of the honey bee's very effective communication system you never have just one robber. They will recruit an army of robbers from their colony and if the colony being robbed is too weak to defend itself, it will very quickly lose all its stores and death of that colony may result.

Robber bees can spread disease, weaken colonies, and cause the colony being robbed to become very defensive and difficult to handle. There are several actions the beekeeper can take to prevent a robbing situation developing, including:

- Good maintenance of equipment – repair any holes or gaps promptly. Be particularly vigilant when you put clearer boards on as the bees that were guarding any holes will no longer be at their stations!
- Maintain strong colonies – we all try and rescue weak colonies, but in reality, this is false economy, and these are the ones most likely to be robbed.

- Don't leave combs out in the open where bees can access them. If there are any stores in these combs they will quickly be robbed out and can create a frenzy that encourages these bees to look for other potential sources.
- Feed bees in the evening after they have stopped flying and try not to spill any syrup, so the bees are not alerted to the possibility of an easy meal.

If a colony has fallen prey to robbers from another colony, then reduce its entrance so it can be guarded more easily. If this doesn't work, your best option is to move the colony being robbed to another site and give it a chance to recover.

↑ There are several different types of clearer board. The rhombus type and Porter bee escapes are shown.

Beware of wasps

The danger does not only come from bees, wasps can also be a real problem in late summer as they switch their diet from protein to sugar. Hives become very attractive to them, and they will force their way in to rob out the honeycombs. Strong colonies will become very defensive and keep the invaders out, but weak colonies will not cope with the wasps' advances and it is not uncommon for a

↑ A simple wasp deterrent made using electrician's conduit. The bees find the entrance points at either end, but the wasps tend not to.

colony to be completely robbed out and then subsequently die from starvation. Watch out for wasps patrolling your hive entrances and sneaking in past the guard bees. If you start to see this activity, it is time to take action. Wasp traps seem a little indiscriminate and there is an argument that they just attract more wasps into the area. They can also trap and kill other flying insects, including our native European hornet, which is a great shame as these are beautiful insects that do not pose a problem to us. There are floors available with entrances underneath, that bees learn to use but wasps do not, as well as other gadgets to reduce the entrance of the hive. If you have a colony that is being severely attacked by wasps, your best option is to move it away to another site.

I have pointed out some of the things that can go wrong, but hopefully this will be a fruitful time of year for you and your bees and you will be able to reap the harvest from your season's efforts. Be organised and tidy in your extracting room, as for many of us it doubles as our kitchen! A webinar on harvesting honey is available on the WBKA website if you would like to see more of the process that is involved. I promise, it is worth all the sticky floors and propolis on the work surfaces when you see your product in a jar for the first time – and I can guarantee that nothing will have ever tasted so good!

8.
The Apiary in August

After the frenzy of activity over the last few months, when there were probably times when it felt like the bees were in control of you rather than the other way around, August should provide an opportunity to enjoy your bees in a slightly more relaxed fashion. There is still plenty going on and you will need to continue weekly inspections to check for swarming, as well as monitoring for disease and checking the performance of queens.

August is the time for heather and the beautifully prized honey that it produces. Use strong colonies with young queens to go to the heather. Nucs produced earlier in the season that have built up to full size colonies will be ideal for this. The queens are young and vigorous and will continue laying late into the season providing a strong foraging force when the heather is at its peak. Look for sheltered sites right in amongst the heather so that the bees do not have far to fly. The weather can be a bit unpredictable at this time of year, particularly in the more upland areas so make it as easy as possible for the bees. If you hit it right with strong colonies and hot August weather, you will be in for a treat.

Weekly inspections

Continue with weekly inspections as it is not uncommon for late swarms to occur, although I have always wondered why bees do this so late in the season. I understand it is a last-ditch attempt to reproduce, but it rarely ends in success this late in the year so surely they would be better going through the winter as a larger colony and putting their efforts into swarming early the next season? But who am I to argue with nature and maybe someone else can offer an explanation?

During inspections have a look at the laying pattern of the queens. Record how many frames have brood on and what that brood pattern looks like i.e. is it an even laying pattern or are there lots of cells that have been missed? How much drone brood is being produced and is this in worker cells? At this time of year, it is really important that you have a healthy, viable queen as she needs to produce the bees that will take the colony through the winter. If you see evidence of a drone laying queen, then it is time to requeen that colony or join them onto another one (having first removed the failing queen). Drone laying queens can be recognised by greater than normal numbers of drones in the colony and drones being produced in worker cells. This happens because the queen is laying an egg to produce a worker but because she has run out of sperm the egg is not fertilised and a drone is the result. The nurse bees extend the cell to accommodate the drone, resulting in worker brood that is raised and more knobbly than normal.

Supersedure

This is also the time of year to look out for signs of supersedure. This is when the colony recognises that their queen needs replacing, and they take steps to produce a new queen while the old one is still in residence. It is a sign of an old, failing queen or maybe a new queen that did not get mated properly and the colony realises she is not viable. Supersedure is characterised by the production of just one or two queen cells typically in the middle of one or two brood frames. When the new queen emerges, it is quite common to have the old queen and her daughter existing side by side and the colony will decide when it is time to despatch with the old queen. If you find supersedure cells in a colony, leave well alone and let the bees manage the situation. Bees do not swarm in a supersedure situation but learning to distinguish between swarming and supersedure may take a little experience. Knowing the history of that queen will help; has her laying rate been slowing down, is there evidence of drone brood dotted around randomly among worker brood and what is her age? Piecing together all this information, which you have been religiously accumulating in your records, will help you understand what is going on.

Varroa control

For those of you that have removed your summer honey crops, now is the ideal time to think about varroa control. The National Bee Unit has a useful leaflet called 'Managing varroa', which provides more detail on this topic. Hopefully, you will have been monitoring the levels of varroa in the colony throughout the summer, either with an inspection board under an open mesh floor or by drone brood uncapping. Both methods will provide an indication of whether the levels of varroa in your colony are reaching a stage that requires treatment. As some of the treatments available cannot be used while honey supers are on the hive, this time of year provides a great opportunity to treat if it is required.

The varroa mites are constantly evolving and some have developed resistance to the synthetic pyrethroids, which were the most popular chemical treatments up until about ten years ago. This highlights the need to be more careful about using the treatments available in a responsible manner, to prolong their lifetime and to ensure we have effective treatments available for use in the future.

The golden rules are:

* Only treat if necessary.
* Always read the instructions on the packaging and use the correct dose. Remove the treatment from the hive after the specified time.
* Do not use the same treatment every time. Alternating with different types of treatments makes it harder for the varroa mites to develop resistance.
* Incorporate Integrated Pest Management (IPM) into your general colony management. Using methods such as a mesh floor, drone brood removal or comb trapping are effective methods to reduce varroa loads.

Some bee breeders are reporting evidence of strains of bees that are more tolerant of the varroa mite. This is great to hear, and these bees provide another important weapon in our armoury against the varroa mite. While many of us do not have bee-breeding programmes, even those of us who are rearing a few queens for our own use should be selecting from colonies that have the lowest varroa levels.

In this book we have glossed over the details of honey extraction because it would take up too much space! However, fear not, the WBKA has some really good resources to help you plan and execute this potentially messy, but very rewarding, activity. On the WBKA website (wbka.com) you will find a booklet called 'Harvesting honey' by Wally Shaw as well as a recorded webinar by Lynfa Davies. Both resources describe the process with plenty of images and descriptions to help you understand what you will need to prepare.

↑ This photo shows two types of heather, *Calluna vulgaris* or Ling heather (the paler more plentiful species in the picture) and *Erica cinerea* or Bell heather.

9.
The Apiary in September

Winter preparations need to begin in earnest now, which means ensuring we have strong, healthy colonies. From the bee's perspective this means the production of winter bees, which are physiologically slightly different to the worker bees produced in the spring and summer. Winter bees have greater fat reserves, which are accumulated by being fed plenty of pollen as larvae and continuing to feed on it as young adults. Pollen is a very valuable and important resource to the colony and should not be underestimated even at this late stage in the season. The winter bees use their fat reserves to survive over the next six months and their parting gift to the colony in the early spring will be the nursing and feeding of the new worker bees that take the colony forward into the new season. You do not want these bees to run out of steam before they are able to raise the new spring workers, which is why they need to be well-nourished during their developmental stages.

Check for disease

The health of the colony is also important at this time of year. Varroa treatments may have been done or are in process, but there are other diseases we can check for. Rather than relying on your seasonal bee inspector to check your hives for disease, why not learn some of the signs so that you can check yourself – after all, you may not have a visit from the inspector every year. As described in April, the first thing to get to grips with is what healthy brood looks like. Open brood should have an even laying pattern with plump, pearly white, segmented larvae in a 'c' shape. Sealed brood should be in slabs with biscuit coloured cappings that are even and dry. Brood that deviates from this should make us take a second look. The diseases we are checking for are

American Foulbrood (AFB) and European Foulbrood (EFB). AFB affects the larvae after it has been sealed and is characterised by dark, greasy, sunken cappings. Beneath these cappings are dead, decomposing larvae. In the later stages dry, hard scales can be found on the lower side of the cells. EFB kills the larvae before they are capped and causes the larvae to appear twisted in their cells and they change colour from white to yellow and through to brown as they decompose. They appear melted as they lose their segmentation. If you suspect either of these diseases, you need to put your apiary on standstill and call your bee inspector.

Other diseases that you might encounter are chalkbrood, particularly in weak colonies, sacbrood which is sometimes confused with EFB and parasitic mite syndrome which is a result of a heavy varroa infestation. Learning how to recognise these diseases is not easy, not least because you are very unlikely to ever encounter some of them. But don't let that stop you having a go and getting in the habit of regularly checking your colonies. Attend your local disease workshops organised by the bee inspectors and make use of the information on BeeBase (www.nationalbeeunit.com) which covers all the diseases and how to treat or control them.

Invasive species

The Himalayan Balsam, *Impatiens glandulifera*, will be in full flower now. This non-native plant is incredibly invasive and, in some areas, clearing of river catchments is taking place to try to eradicate it. Whatever your views on this controversial plant, it does provide a useful nectar source that gives many beekeepers a late season harvest of honey. Ironically, another non-native invasive plant, Japanese knotweed, *Fallopia japonica*, is also highly attractive to insects when it flowers in late summer/early autumn, but I do not think any of us would like to see this aggressive plant become more widespread! Do not forget it is illegal to plant or allow either of these plants to grow in the wild and if you are growing them in your gardens they must not be allowed to spread to the wild.

Keep an eye out for signs of the Asian hornet, *Vespa velutina*. If it gets a foothold in the UK, it will become a significant pest of honeybees and other insects, and we must do everything we can to prevent this happening. At this time of year, hive entrances as well as stands of flowering ivy are the best places to spot the hornets as they prey on bees and other insects. As beekeepers we can play a pivotal role in monitoring as well as educating others in what to look out for.

← Healthy larvae are pearly white, segmented and lie in a C-shaped form.

Feeding

If you have finished taking off your honey crop, it is time to feed the bees if they are not left with enough stores for the winter. As a rough guide each colony will need 20-25kg of stores for the winter. At this time of year, the syrup (either a commercial bee feed or syrup made from sugar and water) can be fed in bulk using a rapid feeder. These sit on top of the brood box or super (if overwintering as brood and a half) and the bees will rapidly empty them and fill their combs. Feed during the evening and take care not to spill syrup in the apiary or you could induce robbing. If you visit the apiary in the evenings while you are feeding you will hear an amazingly loud hum from the hives as they fan their wings on the combs and at the entrance to create air currents to evaporate off the excess water.

At this time of year, you will see the colonies evicting the drones from the hives. Drones are important during the summer but, unfortunately for them, they are too expensive to overwinter and would eat too much of the precious honey stores. You will see them being dragged out of the entrances and refused re-entry. When this happens you really know the season is over.

↑ From left to right: A frame feeder which is used inside the brood box where it hangs in place of one of the brood frames. A float, or small sticks are required to stop the bees drowning. Contact feeders. The white one is available in several sizes. When upturned over the crown board the bees can access the syrup through a gauze cover in the lid. It needs an empty super to accommodate it. The small glass jar is a homemade contact feeder that can be used on nucs or full-sized colonies to provide a small feed. There are small holes punctured in the lid through which the bees access the syrup when it is upturned over the hole in the crown board. The feeder on the right is a large jumbo feeder or rapid feeder. Larger quantities of syrup can be supplied in one go and the bees access it by climbing up through the holes that are covered by plastic cups to prevent the bees drowning. For more detailed information on feeding bees refer to Wally Shaw's booklet on the WBKA website (wbka.com) called Feeding Bees.

The Apiary in October

By now, regular inspections will have ceased and varroa treatments will be complete. We need to prioritise checking that the colonies have enough stores to last through the winter. Bees may still be actively foraging on Himalayan balsam and ivy, if weather permits, adding some additional stocks to their reserves.

Any remnants from varroa treatments need to be removed from the hives. Strips impregnated with chemicals must be removed from the colonies according to the instructions. Leaving them in the hive beyond the specified period will provide a continuous low dose of the chemical, which facilitates the development of resistance in the mites.

Feeding also needs to be completed this month. If you are feeding a syrup made from sugar and water, the bees need to ripen this and remove the excess water. This takes a considerable amount of effort and requires warm temperatures to evaporate the water. So, if you are making your own syrup, make sure the feeding is finished while conditions are mild enough to allow them to ripen it. The consequence of unripened stores is fermentation in the combs and dysentery in the bees. If you are using a manufactured bee feed, you can afford to feed this slightly later into the season as it is presented to the bees in a form they can readily digest. I would aim to have all feeding finished by the end of October in case a cold snap occurs in November. This will cause the bees to cluster and ignore any syrup in the feeders, and you may end up with colonies with insufficient stores to last through the winter.

Now is the time to remove queen excluders, particularly if you are overwintering on brood and a half or double brood configurations. During the winter the cluster of bees will steadily migrate around the boxes to make use of the stores. It is essential that the queen can go wherever the cluster goes – it would be a disaster if she became trapped away from the cluster. Take the opportunity to clean the queen excluder, removing any brace comb ready for next spring.

Prepare for winter

It is also worth checking the condition of hives before the winter weather arrives. If you find holes in, or damage to, any of the boxes, transfer the bees to a new box and take the damaged one to the workshop for repair. Corners of boxes can suffer from overzealous hive tool use during inspections, and this is an area where damage commonly occurs. Rooves must also be weather tight and can often come apart on the corners.

Protection against winter pests should also be applied this month. Mouse guards are fitted at the entrance to prevent these little rodents sneaking in and setting up home in the warm, food-filled haven. During the summer the colony will actively defend itself against these pests, but in the winter when the bees are clustered there is no-one defending the front entrance. The mice will chew the combs eating any bits of brood or honey that they fancy and at the same time doing irreparable damage to the combs and

↑ This hive is protected against woodpeckers with chicken wire. It also has a mouse guard fitted at the entrance and there is a strap in place ready for the winter storms. You will also notice that this colony is configured with the super below the brood box. This super is full of stores and by placing it below the brood box the bees move the stores up to where they need it but more importantly, in the spring, the queen is less likely to be laying in the super. The new brood nest is more likely to be at the top of the stack of boxes. In the spring, the empty super can be removed and placed above a queen excluder on top of the brood box ready to be filled again.

sometimes the woodwork of the frames too. Green woodpeckers can also be a nuisance and they will drill through the side of the hive to access the bees and brood inside. They can drill through wooden boxes with ease and polystyrene boxes provide no challenge at all! Some people report living side by side with these beautiful birds and never experiencing a problem, but it is not worth taking the risk as in harsh weather they will look for alternative feed supplies and once they learn how to access your hives, they will make light work of them. The best defence is to make a wire cage to put over the hive. This makes it difficult for the woodpeckers to get a foothold on the hive and thus prevents them breaking in.

Honey storage

After a bountiful summer season, you need to consider how best to store honey that you have harvested to keep it in the best possible condition and prevent any spoilage. We have all heard tales of how honey lasts indefinitely and has been found perfectly edible after thousands of years in the Egyptian pyramids. I am not sure who volunteered to try it, but I doubt it was their first choice for spreading on their toast!!

↓ Refractometers are available relatively cheaply via the internet and provide a useful check of the moisture content of your honey prior to storage.

Honey is not difficult to store but it does need a little care, and these are the main aspects to consider:

Moisture – honey needs to be below 20% moisture content (ideally 17%) to prevent it fermenting. Fermentation occurs in the presence of moisture, a suitable temperature (18-21°C) and naturally occurring yeasts. The products of fermentation are carbon dioxide and alcohol creating a product that bubbles out of its original container and is no longer fit for sale. Ensure the honey is at the correct moisture content at the point of extraction. Honey that has been capped by the bees in the frames is usually below 20% moisture and any that has not been capped can be checked with a refractometer or a simple flick test. Hold the frame over the uncapping tray and flick it with a sharp jerking motion. If the honey is not ripe it will fly out of the combs and if extracted, it must be kept separate and used immediately.

↑ **Store honey in airtight buckets in a cool room.**

Airtight containers – store the honey in airtight, food-grade containers. Honey is hygroscopic, meaning that it will absorb water from the atmosphere. This can make honey that was below 20% moisture at the point it was put in the container change to become something that deteriorates and ferments.

Temperature – exposure to high temperatures will spoil honey and it should be stored in cool conditions (10°C) to prevent fermentation. Don't be surprised if your honey granulates in the bucket or jars, this is a natural process that occurs most readily around 14°C. The granulated state can be reversed by gentle warming and hence this will be required when you want to bottle your buckets of honey.

11.
The Apiary in November

November is a quiet time for the bees. There may still be some late season forage available such as ivy, which they will take advantage of on mild sunny days, but otherwise the colony is quiet. Brood rearing will have practically ceased making the demands on the colony much less and activity levels will be low. Cleansing flights will take place when conditions allow, but apart from that, the colony appears to do very little.

Winter bees

The colony began its preparations for the winter back in August and September with the production of 'winter bees'. The ability of these bees to survive for five or six months is critical to the survival of the colony as it is these bees that kick start the colony into action in early spring. These bees need to survive the winter months and begin brood rearing as early as late January. The colony cannot afford for these bees to die before the new workers have emerged to take over from them.

So, what is different about winter bees? Normally, when new worker bees emerge in the brood nest the first thing they do is eat. They make their way to the nectar and pollen stores that are located around the edge of the brood nest and after a quick sugar burst, they head for the pollen. The fats and proteins in the pollen help their hypopharyngeal glands and fat bodies develop. By day five they have well developed glands that in a brood rearing colony would enable them to produce the brood food and royal jelly required for rearing larvae. However, by August the colony is producing less brood and so the demand from these young worker bees for brood food is much reduced, so they store

↑ Bees require a variety of pollen in late summer to boost the production of internal fat stores. The more colours you see in the combs the greater the variety.

the fats and protein in their hypopharyngeal glands and their fat bodies. It is this ability to store protein in this way that is the secret to their longevity. Perhaps this is the elixir of youth that we are all searching for?!

As well as having their 'on-board' food stores the winter bees have much lower metabolic rates because they are not actually doing very much. When the ambient temperature drops below 18°C, the bees will start to form small clusters and as the temperature drops even lower these clusters join together. At 0°C the cluster is as tight as it will get with an outer layer of bees tightly holding it all together. In the middle the cluster is not so tight, and the bees are able to move around and feed and, if necessary, vibrate their wing muscles to generate heat. They regularly swap places with each other to ensure they all have access to feed. As there is no brood rearing the colony can afford for the brood nest temperature to drop, and they will maintain the centre of the cluster at about 20°C. It can go as low as 13°C but any lower will be fatal to the bees in the outer shell who lose their ability to cling on to the cluster at 8°C.

So apart from flights on milder, sunny days you can rest assured that the lack of activity is actually helping your colonies to survive. You can make yourself feel useful by checking that autumn storms have not wreaked havoc and I always recommend strapping hives down for extra security.

Reflect on the season

Now is a time for reflection while the season's activities, successes and failures are still relatively fresh in your mind. Do you think your swarm control methods were successful? Were you able to spot the signs early enough and take action before swarms emerged, or was it more of a panic every time you set foot in the apiary? Have a think about where in the process things were not quite working out and take the time to read up on different methods, to see if they will suit your style of beekeeping better.

Did you have a go at queen rearing this year? What was your success rate like? Did mating success vary between apiaries? Consider where the problems arose so that you can pay extra attention to those areas next year. We cannot always blame the weather for poor results when queen rearing, and often it may just been down to poor larvae selection when grafting, or maybe the mating nucs needed to be stronger.

Time to learn new skills

There is always something new to learn in beekeeping and many associated crafts to try out. That is what makes keeping bees so rewarding and I would strongly urge everyone to try out something new. There are plenty of courses available through a variety of providers, from technical topics such as queen rearing, right through to new crafts such as making cosmetics. You may uncover some hidden skills and who knows where it might lead!

↑ Keep a look out for suitable training courses. These are a great way to learn new skills and meet other beekeepers who are interested in similar things.

← Learning a new skill such as queen rearing is rewarding and provides you with all the queens you will ever need.

Late foraging

Foraging on ivy can continue late into the autumn. The honey produced from ivy has a strong flavour that is not to everyone's liking and most people leave it with the bees. Ivy honey granulates to produce hard crystals that some say the bees have trouble using in the winter and there are suggestions that this can lead to starvation. This is not something I have experienced so I am always happy to see the bees making the most of this late forage source. Many other insects are equally happy to forage on the ivy flowers and this makes it an ideal location to keep watch for Asian hornets. They will hawk along areas of ivy predating on all the insects feeding there.

↑ Foraging on ivy can continue late into the autumn.

12.
The Apiary in December

Cold winter days mean there is little activity in the apiary, but you can still delight in seeing your bees flying on the milder, sunny days. Now is a good time to check the apiary hardware to see if it needs replacing. By this I particularly mean hive stands, which may be showing signs of wear. While the bees are not active you can remove them from the stands and make any necessary repairs before replacing them. Also check the hive stands are level and rectify any issues. Moles and rabbits can often cause problems with their digging, making the hive stands unstable.

→ **Check hive hardware regularly to make sure everything is weather proof.**

Winter varroa treatments

It may seem that there is nothing to do in the apiary this month, but now is the best time to treat your bees for varroa infestations. When the colony is broodless, all the mites will be on the bees rather than hiding in sealed cells, so any treatment will be far more effective. The treatment to use at this time of year is an approved oxalic acid based treatment, either using the trickle method or the sublimation method. The research group at the Laboratory of Apiculture and Social Insects (LASI), University of Sussex have done some very thorough work on this where they have investigated both these methods to see which is most effective (www.sussex.ac.uk/lasi/sussexplan/varroamites). They found that oxalic acid cannot penetrate sealed cappings so these treatments need to

↓ Using the sublimation method to treat the colony for varroa.

↑ Inserting board with vapouriser and oxalic acid crystals under floor of hive.

be carried out during a broodless period. Surprisingly, it can be very difficult to find a broodless period and you may need to scrape open any small patches of brood that are found in the centre of the brood nest. It feels wrong to be opening our colonies in December to check for sealed brood, but the LASI team found that it had no adverse effects on the colony and that the operation could be performed very quickly.

Using the sublimation method was found to be the most effective with 97.6% of mites killed compared to just over 93% with the trickle method. With the sublimation method, if a further treatment is given two weeks later the percentage of mites killed is nearly 100. Furthermore, they found that when this level of control is achieved, the duration of control lasts for over one year. The implications for this are significant because if this method of control is applied correctly you will only need to treat your bees once a year, which in my eyes is preferable, as it not only costs less but also reduces the amount of chemicals applied to the colony. It is also useful when taking bees to the heather because you do not need to worry about trying to squeeze a treatment in when they return, instead it can be done in December.

Oxalic acid is a hazardous chemical and should be used with the correct personal protective equipment and care. This applies whether you are using the trickle method or the sublimation method. With the sublimation method the hazard comes from inhalation of the vapour and with the trickle method care needs to be taken when handling the acid.

Christmas presents

The most exciting thing about December is Christmas! I find leaving strategic 'post-it' notes around the house works very well for ensuring the tree has a couple of welcome gifts under it! There is no end to the array of new gadgets available to beekeepers, some more useful than others, and what could be more fun in the depths of winter than being reminded of the new season to come? Books have also made a regular appearance on my 'post-it' notes. There are plenty to choose from so my suggestion would be to think of an area that you would like to improve on and have a look to see what books are available. For example, there are some good books available on honeybee pests and diseases with different titles aimed at the different experience levels.

You may like to brush up on your knowledge over the winter by reading your new books and why not have a go at preparing for one of the assessments or modules available, for example through the BBKA. If this is to be your first module, I would recommend doing Module 1. The BBKA modules do not have to be done in order, but Module 1 follows on nicely from the Basic assessment and most of the information required will be familiar to you if you have kept bees for one or two seasons. Many people shy away from the assessments saying that exams are not for them, but I can honestly say that the whole process is incredibly rewarding and if you work your way through one or more of the modules you will learn so much new information that will really benefit your beekeeping. None of us like the pressure of exams but think of the wider picture and enjoy the process of learning more about bees. More information can be found on the BBKA website and it is also worth asking in your association if there is any support available because some have study groups specifically aimed at helping people through the assessments.

About the author

Lynfa Davies, Master Beekeeper, NDB

Lynfa has kept bees with her husband, Rob, since 2005. During this time she has held a variety of roles with the Welsh Beekeepers' Association (WBKA) including General Secretary and Exam Secretary. She is currently a member of the Learning & Development Committee where she inputs into the development and delivery of courses and workshops for beekeepers across Wales.

Learning new skills has always motivated Lynfa. Once she realised there were a range of opportunities to develop her beekeeping skills, she embarked on the path to acquire beekeeping qualifications through the WBKA, BBKA and the National Diploma in Beekeeping (NDB). Lynfa remembers clearly how daunted she felt when she took her Basic Assessment under the watchful eye of the late Pam Gregory. It proved a pivotal moment, and with huge encouragement from Pam, Lynfa worked her way through the exams and assessments to become a Master Beekeeper in 2015. But it didn't end there as she went on to study for the NDB which she was awarded in 2019. Her journey highlighted to her the importance of access to quality training courses, without which, she believes that her success would not have been possible.

Currently, Lynfa keeps approximately 30 hives which she manages for honey production. She rears all her own queens which, apart from being very cost effective, provides a huge amount of pleasure. Her bees are what she would describe as locally adapted, which she considers important in an area which can be quite challenging due to the high rainfall.

While beekeeping certainly takes up plenty of her time, Lynfa also enjoys cycling, walking, gardening, and generally being outdoors watching wildlife. Her garden is very unkempt, which she justifies because of its value to nature and her bees!

Lynfa acknowledges how, since becoming a beekeeper, her interest in plants and other pollinators has developed. Only bee friendly plants get planted in the garden and she is even known to leave leeks in the vegetable patch go to flower because the bees love them!